FUN WITH LIGHT

by MAE and IRA FREEMAN

KAYE & WARD LTD., LONDON

Photograph credits: Wide World Photos, Inc., pages 3 (lower right), 17 (bottom); Radio Engineering Laboratories, Inc., 21 (bottom); Ballistics Research Laboratory, Aberdeen Proving Grounds, 27 (upper right); Bausch and Lomb, Inc., pages 39 (bottom) and 53 (lower left); Chemoptics Company, for material supplied, 51 (lower right); Jarrell-Ash Company, 57 (lower right).

First published in Great Britain by
Kaye & Ward Ltd
21 New Street, London EC2M 4NT
1968
Reprint (with corrections) 1976

ISBN 0 7182 0070 5

Printed in Great Britain by
Fletcher & Son Ltd, Norwich

CONTENTS

FINDING OUT ABOUT LIGHT

Light tells you almost everything you know about what is happening around you. You are reading this book by light that comes from the page to your eyes. Light lets you know where things are and what is going on. You depend on light for almost everything you do. Light from the sun warms the earth and makes life possible.

Light travels in the form of waves that spread out very much like waves on water. Science knows that light waves can move even through empty space, because they come to the earth from the sun and from the stars.

It is hard to imagine how fast light waves travel. In space, they travel at approximately 300,000 kilometres a second. This amounts to over 17 million kilometres each minute! Light waves can go all the way round the earth in the time it takes you to blink your eye.

Light waves can go through clear materials such as glass, water and air. But the waves do not go quite as fast in such materials as they do in space. In fact, *nothing* can move as fast as light waves travelling through space.

Anything that sends out light is called a SOURCE of light. Outdoors, you get light from the sun and the stars. Light comes from fire, lightning, and even from fireflies and glow-worms and from rotting leaves. Indoors, various kinds of lamps are used to furnish light. A clock with a luminous dial and the TV tube also give off light waves.

Many of the things you see do not give off light of their own. Instead, they throw back some of the light that comes to them from lamps or from the sun. When this light goes into your eye, you see the object. That is how you see this book and most of the other things in the room. You can see the moon and the planets even though they are not light sources themselves. Light from the sun falls on them and some of it bounces off to you.

In this book you will find out how to do some experiments that show many of the facts science has found out about light. Most of the things you need for these experiments can be found around the house.

Read each experiment all the way through. Then get together everything you will need and go ahead with your project.

LIGHT MOVES IN STRAIGHT LINES

One of the most important things about light is that it goes straight from one place to another. Except in very special cases, light does not go round things to get to you.

Here is a good way to show that light moves in a straight line. Get a piece of string a few metres (or yards) long and tie one end to a tree or post. Hold the other end of the string to your eye and back away until the string is stretched tight.

Now look along the string and you can see the tree. Look in any other direction and you no longer see it. This proves that the light by which you see the tree comes to your eye along the straight path of the stretched string.

You can line up a set of toy soldiers to see if a board is straight just by sighting. If light did not travel in straight lines you could not do this.

Look at the beam from a big searchlight. It goes straight out from the source, never changing its direction as it travels through the air.

You are off to one side of the beam, so that light that comes to your eye is not direct light from the source. It is part of the light that is scattered your way by dust or moisture in the air. The direct light goes straight on.

A beam of light is just a path in which the light waves are travelling. If the beam is made very narrow, it becomes a light RAY. A ray of light is any line that can be drawn outward from the light source to show which way the light is travelling.

As you read through this book, you will see how scientists use rays to trace what happens to light as it goes from one place to another.

Look along the string to see the tree.

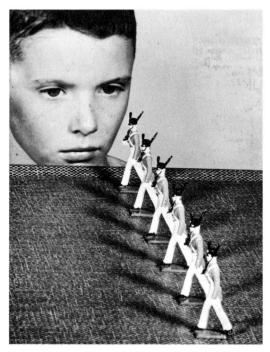

Line them up by looking.

Each beam is a straight line.

PICTURES FROM A PINHOLE

Use a sharp pencil or nail to make a clean hole near the centre of a piece of cardboard, as in the picture. The hole should be about 3 mm ($\frac{1}{8}$ in.) across. Make the edges clean and sharp by rolling the pencil round in the hole.

Tape the card to a block of wood so that it stands up. Darken the room and put a lighted candle a short distance in front of the hole. Then hold up a white card on the other side of the hole. You will see a weak, upside-down light-picture of the candle flame on the card. A light-picture of this kind is called an IMAGE.

This arrangement is sometimes called a PINHOLE CAMERA. It works like this: light rays from each tiny place in the flame go through the hole and fall at a certain place on the white card, making a little patch of light there. This happens for every part of the flame, each one making its own light spot on the card. The whole set of spots forms the complete image. It is a little fuzzy-looking because the light spots are not sharp points and they overlap a little.

Try moving the white card slowly back and forth. Bringing it closer to the hole makes the image smaller, sharper and brighter. Later, you will see how to use a lens instead of a pinhole to get a better image.

Here is another way to make pinhole images:

Pour some sand or small pebbles into a cardboard box. Switch on a small torch or bicycle lamp and stick it into the sand, facing upwards. Put the cover on the box and darken the room. Punch a hole in the box top with a large needle or medium-sized nail. At once you see a spot of light on the ceiling. This spot is a pinhole image of the torch bulb.

Punch out more holes, one after another, and watch the images appear on the ceiling like a sky full of stars. If you hold a piece of card above the box as in the picture, the images will be sharper and brighter.

A pinhole camera.

Each punched hole gives an image.

LIGHT CAN BE TURNED BACK

If light waves hit a smooth, shiny object, they bounce back again just like water waves hitting the side of a pool. This turning back of waves from a surface is called REFLECTION. You can do an experiment to find out the exact direction reflected light will take.

Hold a small mirror in the sunlight coming into the room from a window. Tilt the mirror in various directions and notice how a bright patch of reflected light moves round on the walls and ceiling of the room.

In the upper picture, rays have been drawn to show the direction of the light coming to the mirror and bouncing away again. The ray coming in from the sun has a certain slant above the mirror. The reflected ray that goes to the bright spot has exactly the same slant on the opposite side.

Another experiment will actually let you see the beams of light before and after reflection.

Cut a hole about 25 mm or 1 in. across near one edge of a large sheet of cardboard. Get a big comb with wide teeth and tape it over the hole. Lay a sheet of paper on a table in direct sunlight. Do the experiment early in the morning or late in the afternoon, when the sun is low.

Hold up the cardboard with the hole in it so that it faces the sun. This lets narrow beams of light come through the teeth of the comb and streak across the paper. Now, if you stand a mirror in any position across these beams, you will see a set of reflected beams on the paper. In this way you can see that the beams have exactly the same slant coming towards the mirror and going away again.

Turn the mirror to a slightly different position. This will change the angle of the two sets of beams, but notice that they still match each other. Instead of the light from the sun you can use a beam of light from a torch in a darkened room.

Your experiments have shown an important scientific fact about the reflection of light: rays hitting a mirror at a certain angle always bounce away at a matching angle.

Save the comb and cardboard for the experiments on page 30.

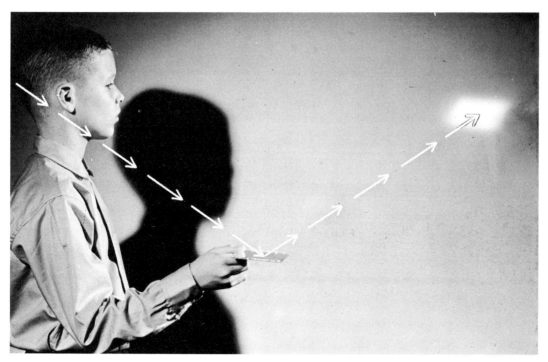

The reflected beam moves when you tilt the mirror.

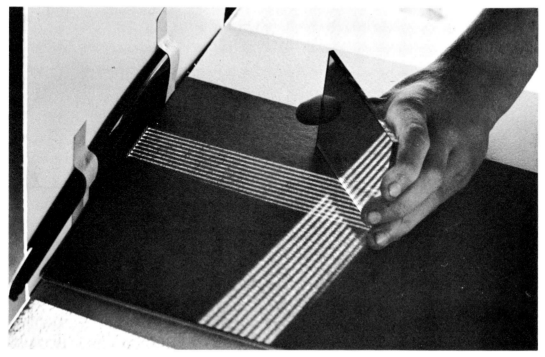

Testing the angle of reflected sunbeams.

GLASS REFLECTS

At night, bright objects in a room are reflected by the glass of the windows. This shows that a piece of glass can reflect light like a mirror, even though it has no coating on the back. The reflection does not show up in the daytime because of the stronger light that comes through from outdoors.

Prop up a pane of window glass between two stacks of books on a table. Stand a lighted candle in a lid and put it down a short distance in front of the glass pane. Now put a tall glass of water on the other side of the pane, just opposite the candle. The glass of water should be the same distance behind the pane as the candle is in front. Darken the room and look at the arrangement from the front. You will see that the candle seems to be burning under water!

What you see in the water is only an image of the candle and flame. It is formed by light that comes from the real candle and bounces off the pane of glass. This image is behind the pane, just where the glass of water is, and so flame and water seem to be in the same place.

You can puzzle your friends even more by doing the experiment another way. Take away the glass of water and put your finger in its place. It will seem as if you are calmly holding your finger in the flame!

The 'reflectors' used on cars and bicycles send back light from many little hollows on the rear side of the glass. This makes the reflector look almost as bright as a lamp.

Modern road signs are often made of a special cloth that has thousands of tiny glass beads glued to it. The headlights of an approaching car make the sign glow very brightly with light reflected inside the beads.

Are both candles real?

A bicycle reflector.

Reflected light makes the signs glow.

REFLECTIONS ON THE INSIDE

The experiment on page 8 showed that light can bounce back from a piece of clear glass. Even when light moves inside a material such as water or glass, it can be reflected when it hits the outer edges of the material. You can do experiments that prove this.

Stand a glass of water at the edge of a table. Look upwards through the side of the glass and see the top surface of the water from below. It looks silvery, almost like a mirror. It _is_ a mirror. Prove this by sticking the pointed end of a pencil down into the water. Looking from below, you see a very good reflection, as in the picture.

Notice, too, that the point and its reflection look much bigger than they really are. That is because the curved side of the glass makes the water act as a magnifier. You will find out about this on page 28.

Get a glass baking dish and hold a lighted torch against one of its edges. The light goes all the way round, bouncing backwards and forwards inside the glass itself, and even going round the corners. Then it comes out again on the opposite edge, as you can see by the bright patch there.

The reflection of light inside a material has many uses. Dentists use a special lamp with a long plastic rod that carries light from a source to the inside of your mouth. The light moves through the curved plastic rod just as water moves through a pipe.

From below, the water surface is a mirror.

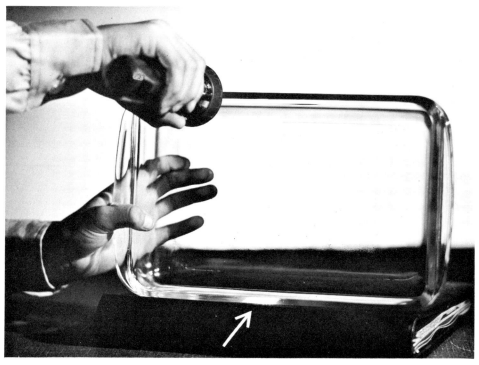

The light comes out on the opposite edge.

FLAT MIRRORS

Print the word LIGHT clearly on a piece of paper. Stand a mirror straight up just behind the word and look at the image. Notice that the letters are turned upside down.

An image in a flat mirror is always turned either top to bottom or side to side, depending on how the mirror is placed. The 'I' and the 'H' seem normal only because these two letters happen to look the same either way.

Lay a piece of carbon paper on the table, with the carbon side up. Put a piece of writing paper down on it and write your name. When you lift the paper you will have your name written backwards on the other side. Hold this writing up to a mirror. The mirror changes the writing round once more so that you see it correctly.

How high must a mirror be for you to be able to see yourself in it from head to toe? Most people would say that the mirror has to be as tall as you are. Here is how to find out.

Ask a friend to stand facing a large mirror that is mounted flat against a wall. Without moving, she tells you where to stick a small strip of tape to mark the place where she sees the top of her head. Put another piece of tape at the place where she sees her toes.

Now take a ruler or tape measure and measure the distance between the marks. You will find that this is exactly half your friend's height.

The drawing shows how the light rays coming from your head and feet are reflected by the mirror to your eye. That is why the only part of the mirror needed for a head-to-toe view is the part between the two tapes. As long as the mirror is straight up and down, you cannot see any more of yourself by backing away or coming closer.

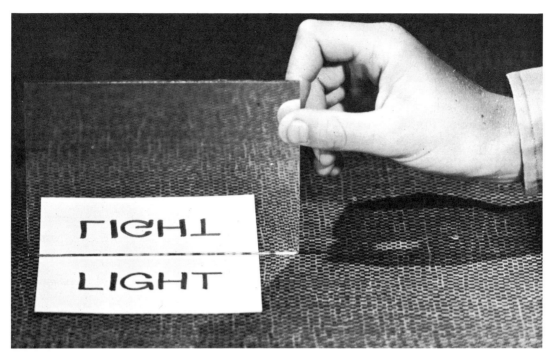

A mirror can turn things round.

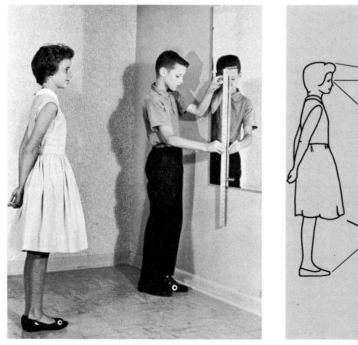 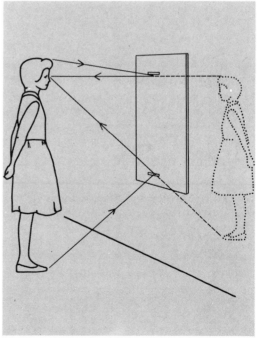

Using light rays to find your height.

MIRROR PAIRS

Get two plain, unframed mirrors and tape them together so they can stand on the table like the covers of a book. Place a toy soldier half way between the mirrors. Now, if you look into the mirrors, you will see a number of soldiers standing round a circle, as in the first picture. You can swing the mirrors closer together or farther apart to get different numbers of soldiers.

Do another experiment with this pair of mirrors. Set them up so they stand at right angles to each other, like two walls of a room. You can do this by laying a book down on the table between the mirrors, with its corner at the place where the mirrors meet. Swing each mirror until it lines up along one edge of the book. Carefully take away the book, leaving the mirrors where they are.

Now look into the pair of mirrors so that you see half of your face on each side of the middle line. Touch your right ear. You see your image touch the other ear!

The image you saw was made by light reflected from two mirrors, one after the other. Each mirror turned the image round once, so finally it was the right way round again. You can understand this if you imagine yourself in the place of your image. Then you see that it is the right ear that is being touched after all.

A single mirror shows how you look with the right and left sides of your face changed round. But the double mirror changes the image round once more and shows you as you really are. This is how you look to other people.

In another experiment, stand one of your mirrors straight up against a block of wood or some books. The mirror should be facing you. Set a toy soldier down about 25 mm or 1 in. in front of the mirror. Then hold the other mirror on the table about the same distance in front of the soldier, so that it faces the first mirror. The top right hand picture shows how to do this.

Look over the top edge of the front mirror and you will see a whole company of soldiers, stretching away into the distance. Adjust the mirror until the row is perfectly straight. All the images are formed by light that is reflected back and forth between the two mirrors.

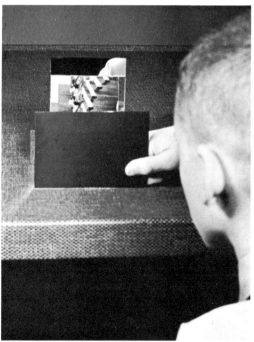

How many soldiers?

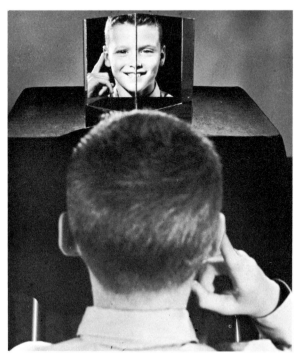

The right ear seems to be the left one.

MAKING A PERISCOPE

Another use for a pair of mirrors is to move a beam of light to one side. That is what happens in a PERISCOPE, which is an instrument that lets you see round corners.

Make a periscope of your own. Get a cardboard box that has ends about 6½ cm or 2½ in. square. A biscuit packet is just right. Cut three of the edges at the top of the box to make a lid that you can lift up. Then cut out two square holes at opposite corners as shown in the picture.

Now fasten two small mirrors inside the box with sticky tape. The mirrors should face towards each other. Be sure to have the two ends of each mirror the same distance from the corner of the box, to form an even triangle. Tape down the lid of the box and your periscope is ready. You can use it to look round a corner without being seen.

On the periscope of a submarine, a small tube carries the upper mirror. The tube is run up when the periscope is to be used and pulled down again afterwards.

Periscopes are also useful in armoured tanks and in space capsules. With the help of a periscope, a scientist in an atomic laboratory can watch the dangerous materials he is working with while sitting safely behind a thick wall that protects him from the harmful rays.

Make your own periscope . . .

. . . and use it to look round corners.

At the periscope of a Polaris submarine.

MIRRORS THAT BULGE OUT

The mirror experiments you have done up to now all used flat mirrors. Curved mirrors are useful and interesting, too. Try some tests with mirrors that are curved outwards, like a ball.

Get a plain, silvered Christmas-tree ball. Look at the reflections in it and notice that you can see the image of nearly everything in the room.

Now bring the ball close to your face and look at your own reflection. Your nose, which is quite near the ball, looks very big. Your ears and chin, which are farther away, look much smaller. In the picture, you can see the image of the camera, just to the left of the bright spot, if you look closely.

Mirrors that curve outwards always change the shape of things that are reflected in them. Another experiment with this kind of mirror will show that you can straighten things out again by using the mirror twice.

Make a simple drawing of a building as shown in the middle of the left-hand picture. Place the Christmas-tree ball on the paper just behind the drawing.

Next, copy the image that you see in the ball on another sheet of paper, as accurately as you can. This image will be very much out of shape. The straight lines of the original picture will all be curved.

Now, put your new drawing down in front of the ball and look at its image there. The image will seem to be straightened out, just like the picture you first made. If the image still looks a little out of shape, move the ball backwards and forwards until it seems right.

Some rear-view mirrors used on bicycles and cars are the curved-out kind. They let you see things far off to both sides of the road. These rear-view mirrors are not curved as much as your Christmas-tree ball, so they do not spoil the shape of reflected objects so much.

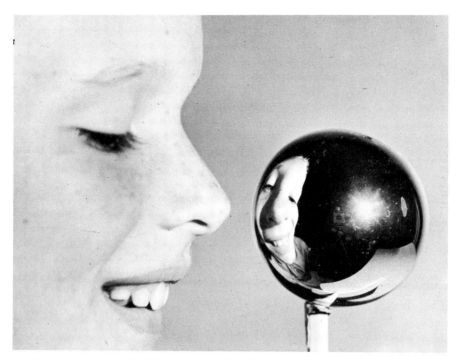

A Christmas-tree ball is a curved mirror.

The reflection is out of shape . . .   **. . . but it can be set right again.**

HOLLOW MIRRORS

In the last experiment, you discovered what curved-out mirrors can do. Now you can make some tests with the opposite kind of mirror—one that is hollow, like a dish. A shaving mirror is a good example. You will find that it works in quite a different way.

Look into a shaving mirror and your face seems quite large. A hollow mirror makes things look bigger, while a curved-out one makes things look smaller.

There is one thing that only *hollow* mirrors can do. They can throw an image on to a surface, such as a wall or a card.

Set the shaving mirror on a table, facing a window. Move a large card around in front of the mirror until you get a sharp image of the window, as in the picture. Notice that this image is upside down. It is also turned round side to side. You can check this easily by standing something on the window sill and looking to see where its image is.

You must choose the shaving mirror carefully for success in this experiment. Many of these hollow or convex mirrors, as the scientist calls them, are so badly shaped that they will not give a good image on a surface. You can test the mirror before purchase by taking a white card to a shop and seeing if you can get an image of a window or doorway.

The bottom picture shows a pair of hollow mirrors set up by the American Air Force in Alaska. Each one is as tall as a 15-storey building. These mirrors are used to reflect radio waves in the same way that your mirror reflected light waves. In one, radio waves from a station hundreds of miles away are gathered up by the mirror and reflected to a receiver at the top of the tower. The other mirror works just the opposite way. It sends out a beam of radio waves the way a searchlight sends out a beam of light.

Move the card until you get a sharp image of the window.

These hollow mirrors reflect radar waves.

GLASS CAN BEND LIGHT

When you look out of the window, you see things outdoors by the light that they reflect. The light waves go first through the air, then through the window glass and again through the air to your eye.

Light waves do not go as fast in glass as they do in air—they go only about $\frac{2}{3}$ as fast. If light waves hit the surface of a piece of glass at an angle, the slowing-up makes them swing aside a little. This change in direction of light waves is called REFRACTION.

Get the glass baking dish that you used to do the experiment on page 10 and put it on the table. Hold a pencil against one side of the dish, as in the picture. Look through the glass in a slanting direction and notice that the pencil appears to be broken. The part of the pencil behind the glass seems to be cut away and pushed to one side. This is because light by which you see the lower part of the pencil is bent aside by refraction as it goes through the glass.

Try another experiment that uses both refraction and reflection. Set a lighted candle on a table and turn off all other lights in the room. Hold a small mirror straight out from your face. One edge of the mirror should be close to the side of your nose, and the opposite edge should point towards the candle. You will see the flame reflected in the mirror, but instead of one image, you will see several side by side.

The diagram shows why there is more than one image. The light rays enter the glass and are then reflected back and forth between the plain and silvered sides. At each reflection, a little of the light gets out and comes to your eye. Each of these rays gives you a separate image, and you may see a whole row of them, each one weaker than the one before it.

The pencil seems to be broken.

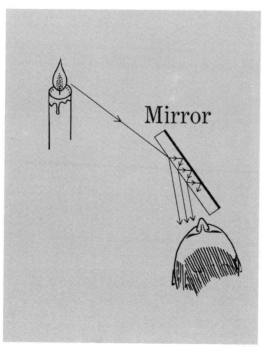

Your eye sees a whole row of candles.

WATER CAN BEND LIGHT

Put a coin in a cup and set it on the table in front of you. Move the cup away from you until the rim just hides the coin.

Without moving your head, slowly pour water into the cup. At a certain point you will be able to see the coin again.

Here is what happened. At first, light from the coin could not get to your eye because the rim of the cup was in the way. But by pouring enough water into the cup, a ray of light coming from the coin could be refracted to your eye, as the upper picture shows.

Have you ever noticed that the water in a swimming pool never looks as deep as it really is? This is because of the refraction of light by the water. You can do an experiment that shows how something under water seems closer to the top than it really is.

Drop a coin into a bucket nearly full of water. The coin helps you to tell where the bottom is. Looking straight down into the water, you can see that the coin seems much closer to you than the floor of the room.

Keep looking into the water and point to the place on the bucket where the coin seems to be. You will find that your finger is only about $\frac{3}{4}$ of the way down from the surface of the water. This can be explained by the fact that light goes only $\frac{3}{4}$ as fast in water as in air.

When you pour the water in, the coin appears again.

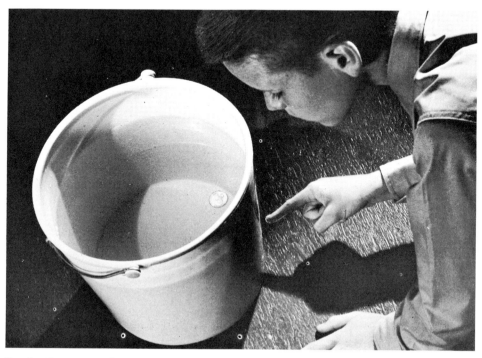

Is the bottom of the bucket really raised?

EVEN AIR CAN BEND LIGHT

Set up a lighted candle on a table about two thirds of a metre (or two feet) away from a wall. Turn off all other lights in the room. Now stand on the other side of the room, switch on your torch, and shine it on the wall. You will see a shadow picture of the warmed air that streams up from the candle flame.

The heated air expands and is then a little thinner than the rest of the air. Light waves coming from the torch are refracted when they go through the warmed air. Instead of going straight to the wall, the waves are bent aside, leaving a weak shadow.

Anything that moves very fast stirs up the air around it. The stirred-up air refracts the light and gives a shadow picture, just as in your experiment. Scientists study these pictures to find out how to make aeroplanes and missiles move more easily through the air.

Refraction of light in the air often plays tricks on you. Sometimes when you are riding in a car on a hot summer's day, watch the road ahead carefully. Suddenly you may see what seems to be a pool of water. But when you come up to that place, you will find that the road is perfectly dry!

Light coming from the sky is refracted by the warm air near the road. The light waves are bent upwards to your eye, as the diagram shows. The shiny pool you see is not water at all, but really refracted daylight.

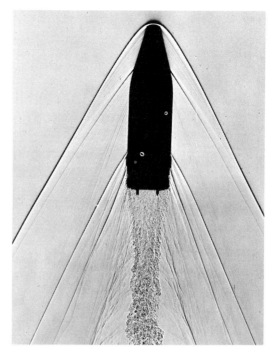

Warmed air can cast a shadow . . .   **. . . and so can air stirred up by a bullet.**

The road is really dry.

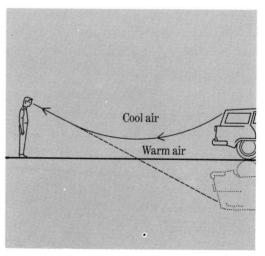

Heated air near the road bends the light and forms an upside-down image of a car.

WHAT A LENS IS

Many hundreds of years ago, people noticed that things seemed bigger when they looked at them through a glass bead or a drop of water. These were probably the first LENSES ever used. A lens is any piece of clear material that has smoothly curved sides.

Nowadays, lenses of many kinds are used in cameras, microscopes and telescopes. Most lenses are made of special kinds of glass. Machines grind and polish the pieces of glass to very exact shapes.

Buy a cheap magnifying glass at Woolworth's. To show how this magnifies things, hold it a few centimetres or inches above the print in a book or newspaper. The letters that you see through the lens look much bigger than the others, as the first picture shows.

The next picture shows how you can get the same result with a lens made of water. Take the curved glass off an old alarm clock. Pour a little water into it, and the curve of the glass will shape the water into a lens. Hold this lens above the printed page and you see the letters enlarged. This water lens does not magnify as much as the glass lens. Also, the letters are a little crooked, especially near the edges. That is because the water lens is not shaped as perfectly as a glass lens.

There is another kind of lens that makes things look smaller instead of bigger.

Borrow some spectacles from a person who is short-sighted. Hold the glasses a short distance above the print and look through one of the lenses. The picture at the lower left shows what happens. If you cannot get spectacles of this kind, use the bottom of a very thick drinking glass, as in the last picture. The letters will look small, but they will be quite badly out of shape because the bottom of the glass is not curved very exactly.

The two kinds of lenses that you tested are different in an important way. Lenses that magnify are *thicker* at the middle than at the edges. Lenses that make things look smaller are shaped just the opposite way: they are *thinner* at the middle than at the edges. Look carefully at the lenses you used in the experiments and you will see that this is so.

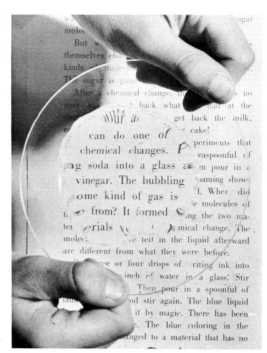

These lenses magnify.

These lenses make things look smaller.

WHAT A LENS DOES

The last experiment showed that there are two kinds of lenses: a curved-out lens makes things look larger, and a hollow lens makes things look smaller.

Each kind of lens works by refracting light in a special way. Find out about this by using the same comb and card that you worked with on page 6.

Set up the comb and card again in the sunlight of early morning or late afternoon, or use a slanting mirror to bring the sun in low, as shown at the top of page 7, or use a torch beam in a darkened room.

Lay another large card on blocks or books a few centimetres or inches from the comb. This card should be level with the middle of the hole in the first card. Adjust the second card carefully until it catches the beams of sunlight coming through the comb. These beams are all PARALLEL. This means that they go ahead in exactly the same direction, keeping the same distance apart.

Now hold the magnifying glass against the edge of the card, as in the upper picture. See how all the beams are brought together to a point. This point is called the FOCUS of the lens.

Bringing light to a focus is one of the most important things a lens can do. In a camera, the lens brings light to a focus on the film to form an image there.

Now do the experiment again, but use the hollow lens you worked with on page 28 in place of the magnifying lens. This time the beams of light are spread out instead of coming together to a focus.

When hollow lenses and curved-out lenses are put together in certain ways, they can do many things with rays of light. That is why groups of lenses are used in cameras, binoculars, telescopes, microscopes and other instruments.

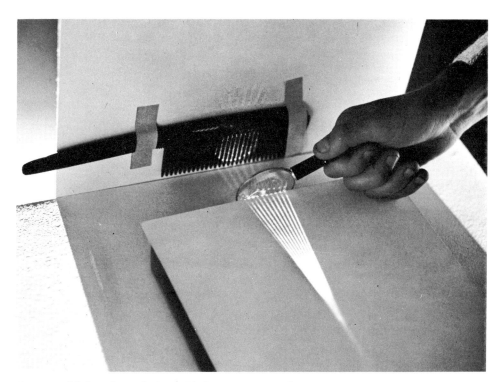

A magnifying lens brings light rays together.

A hollow lens spreads light rays apart.

USING LENSES

Set up a lamp so that the bulb faces upwards. Light the lamp and hold your magnifying glass level, a few centimetres or inches above the bulb. Move the lens slowly up and down until you find the place where you get a sharp image on the ceiling. The printing on the end of the bulb will show up large and clear. The letters in the image will be upside down and turned round side to side.

What you have just set-up is really a home-made PROJECTOR. The bulb is the source of light, the printing is the slide and the ceiling is the screen. Without the lens, the light waves from each place on the bulb would spread all over the ceiling. The lens brings them together to form an image.

Stretch a piece of string out along the floor and tape it down tightly at both ends. Stand on the string at one end and look down at your feet through a pair of binoculars or an opera glass. You must hold the binoculars the wrong way round, with the large lenses to your eyes. Turn the screw until your feet are in sharp focus. How far away your feet seem!

Keep looking through the binoculars as you try to walk along the string. Take short steps and put one foot directly in front of the other. Also, try walking up the stairs, but be ready to grab the rail!

It was hard to keep your balance because you were not looking directly at your feet but at a far-away *image* of them. Each side of the binoculars has a set of two lenses in it. There is a magnifying lens at the large end and a hollow lens at the other. When you look through this pair of lenses the right way, you get a large image, which makes things look nearer. But when you look through them the other way round, you get a small image, making things look far away.

A home-made projector.

Try to walk a straight line.

A FACE IN SPACE

Put a slide into a projector and point the projector towards a far corner of the room. Stand a large piece of cardboard on a chair about 2 metres (or 6 feet) away from the projector and facing towards it. Focus the lens until you get a sharp image on the cardboard. Then take away the cardboard and chair.

Now get a thin, flat piece of wood, such as a long ruler. Hold it where the card was and swing it rapidly up and down. As you do this, you see the image of the slide right there in the air!

Here is what happens. The ruler picks up a strip of the image at each place. Your view of each strip does not fade out as quickly as the ruler moves on. Instead, it lasts in your eye for about a tenth of a second. As the ruler moves, you are able to see all the rest of the image before the other parts fade away.

The same thing happens when you look at a film or television. Each picture flashes on the screen for only a small part of a second. But it does not fade from your eye at once, so you get the idea of smooth motion.

MAKING A MICROSCOPE

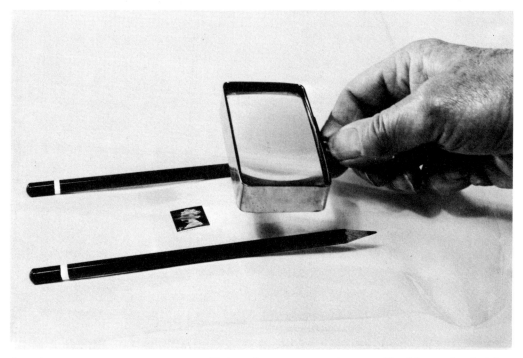

A microscope uses two magnifying lenses to make small objects look big. You can find out how it does this by setting up a simple kind of microscope.

Put two pencils on the table roughly 4 cm or $1\frac{1}{2}$ in. apart. Place a postage stamp between the pencils. Lay a small piece of polythene or cellophane across the pencils.

Dip a drinking straw into water and carefully place one drop on the polythene right over the stamp. Keep the polythene stretched so the drop is about 6 mm or $\frac{1}{4}$ in. above the stamp. When you look through the drop, you see an enlarged image of part of the stamp. That is because the drop of water is rounded like a lens.

This is only an ordinary magnifier. Now make it into a microscope, which will make things look even larger: hold your magnifying glass over the drop and look down through both of them, as in the picture. Keep the glass level and move it slowly up and down until you see a sharp image of the stamp. Be sure you are looking through both lenses. The final image will be greatly magnified. It will also be upside down and turned around from side to side.

The water drop, which takes the place of the front lens of an ordinary microscope, forms an image a short distance away. Then the glass lens magnifies this image still more. Some microscopes can magnify things thousands of times.

HOW BRIGHT IS LIGHT?

Sufficient light must shine on an object before you can see it properly or take a picture of it. The strength of a beam of light can be measured.

Get two blocks of paraffin wax at a chemist's shop. Cut a piece of aluminium foil the same size as the flat side of each block. Lay the foil between the blocks and hold the 'sandwich' together with two rubber bands.

Light a small birthday candle and set it up in a dish near one end of a table. Set four more candles in a dish near the opposite end of the table. Stand the pair of blocks somewhere between the two light sources. The flat sides of the blocks must face the candles squarely. Turn out all other lights.

Now step back and look at the edges of the blocks. One will glow brighter than the other, as in the picture. Move the pair of blocks between the two sources until you find a place where both blocks look equally bright. You will find this place to be much farther from the strong source and closer to the weak one.

The blocks bought at the chemist's might not be shaped well enough for success, but you can easily make the blocks. Put the wax into a tin and melt it by standing the tin in hot water. DO NOT HEAT THE TIN DIRECTLY, the wax might catch fire. When the wax is melted pour the liquid into two shallow cardboard lids and leave it to solidify. The blocks must be of the same thickness. Take care also to keep the aluminium foil free from any crinkles. Part of the light that hits each block is scattered sideways through the edge. When the edges match in brightness, you know that the strength of the light from the two sources is the same. Scientists use this kind of experiment to compare the strength of light sources.

BLACKER THAN BLACK

Whenever light strikes an object, only part of the light is reflected. The rest is absorbed into the material. Dark-coloured, rough materials do not reflect as much as light-coloured, smooth ones.

A black surface reflects almost none of the light that falls on it. But even an ink spot or a piece of black cloth still reflects a little light. The only thing that can be *really* black is a hole!

Cut a hole about 12 mm (or $\frac{1}{2}$ in.) across in one end of a shoe box. Spread black liquid shoe polish or black ink all over this end of the box. When it has dried, the painted surface will look quite black. But the hole seems to be a much deeper black. It is the blackest thing you can possibly get.

This is why the hole looks so dark; any light that goes into the hole will bounce around from one side of the box to another, again and again. Each time the light is reflected from a side, a little of it is taken up until, after many reflections, there is practically nothing left. No light gets out again because the hole traps all of it. The hole is perfectly black.

Look at your eye in a mirror. The little black dot at the centre is the PUPIL—the opening where light enters. Now do you know why it looks black?

THE GHOST OF SUNLIGHT

Get a piece of glass that has a slanted edge. You may find a glass ashtray or vase that has an edge of this kind. A crystal lamp pendant that you can buy at a lighting-fixture shop would be best of all. It should be the shape that is called a PRISM. A prism is any piece of glass with flat sides that come together to form a sharp edge.

Hold the prism in direct sunlight outdoors where there is some shade nearby. Put a piece of white paper on the ground in the shade. Now turn the prism slowly with your fingers until you see a bright rainbow of colours fall on the paper.

You have just repeated a famous experiment that was first done by the great scientist Isaac Newton about 300 years ago. He called the rainbow-like band of colours a SPECTRUM, which means ghost. Each colour of the spectrum blends softly into the next one—red, orange, yellow, green, blue and violet.

Where do these colours come from? Newton showed that they are all in the beam of sunlight, but do not show up as colours until they are separated out.

Sunlight is refracted when it goes through a prism. But each colour is refracted a little differently. As a result, the colours spread out so that you can see each one by itself.

As the light waves travel through the air, they stay an even distance apart. This distance is called the WAVE LENGTH. Ripples on water may have a wave length of a few centimetres, but light waves are very much shorter.

Each colour of light has a different wave length from all the others. Red has the longest waves of any light. Even so, it takes nearly 14,000 of them to stretch a distance of just one centimetre. Violet light has the shortest wave length. There are over 26,000 to a centimetre. All the other colours have wave lengths that fall in between.

Separating the colours in sunlight.

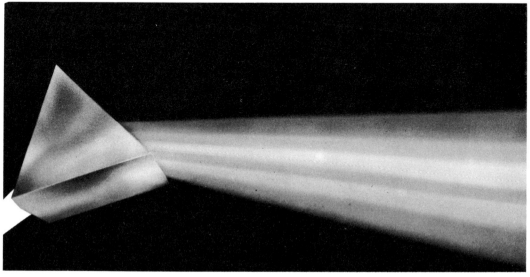

The rays fan out to show the colours.

A WATER PRISM

You can use a prism made of water to break up sunlight into a spectrum:

Use the glass baking dish from the experiment on page 22. Pour water into the dish nearly to the top and set it in the sun near a window. Cut a hole about 25 mm (or 1 in.) wide and 12 mm (or $\frac{1}{2}$ in.) high in a piece of aluminium foil and wrap it around a small mirror, as shown in the lower picture.

Lay the mirror in the water at a slant, with the upper end resting against the side of the dish. Put a stone in the dish to keep the mirror from sliding down.

Now hold up a large card facing the dish. Move it around until you catch a bright spectrum on it, as in the upper picture.

The prism that forms this spectrum is the wedge of water between the mirror and the water surface in the dish. Sunlight goes down through this prism and back up again. This separates the colours from each other, as the drawing shows, and you can catch the spectrum on the card.

Now that you have formed a spectrum, do some tests with it.

Get some pieces of glass or cellophane of various colours. Prop up the card so that your hands will be free. Hold a piece of the cellophane somewhere between the mirror and the card. Watch the spectrum as you do this and see how the cellophane lets through only its own colour and holds back most of the other colours. Check this by trying the different coloured pieces of cellophane, one after another.

Do another experiment with your spectrum. Put a magnifying lens into the path of the light that forms the spectrum. Hold the lens about 7 to 10 cm (or 3 to 4 in.) from the card. In place of the spectrum you will now see an image of the hole in the foil. You may have to move the lens back and forth slightly to get a sharp image. You will find this patch of light has no colour at all.

The lens brings all the spread-out colours together and the mixture again looks like sunlight. This proves that sunlight is really all the colours of the spectrum put together.

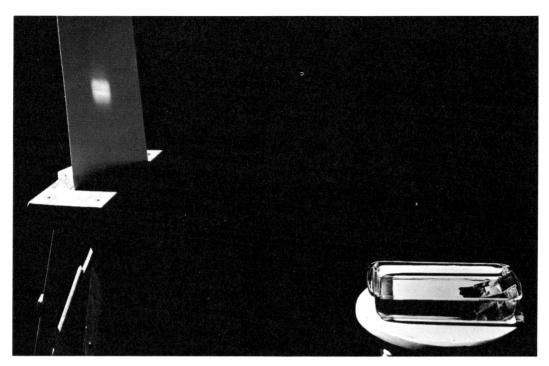

Water and a mirror make a prism.

Wrap the foil around the mirror.

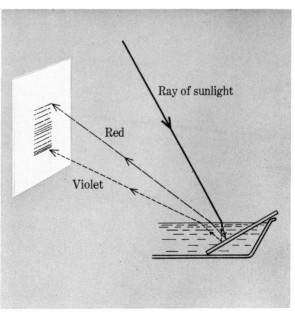

Ray of sunlight

Red

Violet

The light rays go into the water and out again.

RAINBOWS

There is a beautiful spectrum that people were always able to see even before they knew about prisms. It is the rainbow that appears in the sky after a storm.

When the rain has stopped falling, there are still tiny drops of water that keep drifting around in the air. As the sunlight goes through these drops, the rays for the different colours are bent in different amounts, just as in a prism. As a result, you see a huge spectrum in the sky. The sketch shows what happens to the rays as they go through a raindrop.

Make your own rainbow by using a garden hose. Do this in the early morning or late afternoon, when the sun is low. Stand with your back to the sun. Aim the stream of water upwards and adjust the nozzle to give the finest possible spray. Move around a little until you can see the rainbow in the spray. It is best to look at it against a dark background, such as some trees.

Try another experiment to show how even a single drop can form a rainbow.

Get a clear, uncoloured glass marble and mount it on the flat head of a nail with a drop of glue. Cut a clean, round hole in the centre of a large card. The hole should be a little bigger than the marble.

Face the card towards the sun and hold the marble about 5 to 8 cm (or 2 to 3 in.) behind the card, in line with the hole. Move the marble around slightly until you get a circle of light on the card. The bright edge of this circle is a rainbow, but it is very narrow and you must look closely to see the colours.

Look for the rainbow in the spray.

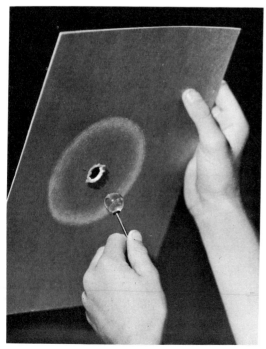

A glass marble makes a rainbow.

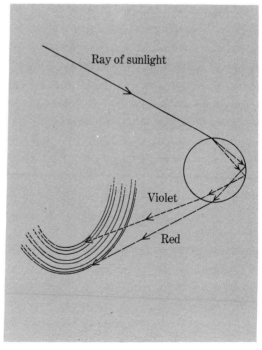

Ray of sunlight

Violet

Red

Light rays in a raindrop.

SPIN A COLOUR TOP

In one of the experiments on page 40, you used a lens to bring together all the colours of the spectrum. This made a spot of light that showed no colour at all. You can get the same result by using a spinning top instead of a lens.

Cut a disc about 82 mm (or $3\frac{1}{4}$ in.) across out of stiff cardboard. Make a pencil mark on the edge of the disc and hold it upright on the table with the mark at the bottom. Lay a ruler down just in front of the disc. Be sure that the left-hand end of the ruler is exactly at the mark. Carefully roll the disc along the ruler without slipping and make a mark at each of these places: 76 mm (3 in.), 114 mm ($4\frac{1}{2}$ in.), 146 mm ($5\frac{3}{4}$ in.), 190 mm ($7\frac{1}{2}$ in.), and 234 mm ($9\frac{1}{4}$ in.). The picture shows how to do this.

Draw a straight line from the centre of the circle to each mark. Then paste wedge-shaped pieces of coloured paper over each section. Use red for the largest section, then follow to the right with orange, yellow, green, blue and violet. Be sure you use the colours in exactly this order.

Make a hole in the centre of the disc with a needle. Sharpen one end of a piece of wooden matchstick and push it straight into the hole, letting only about 5 mm (or $\frac{1}{4}$ in.) come through on the other side. The finished disc should look like the one in the lower-left-hand picture.

Now, with a snap of your fingers, set the disc spinning like a top. The colours disappear and the disc looks white!

Each section of the disc reflects one of the colours of the spectrum. These colour images last for a moment in your eye and pile up to make the whole disc look white.

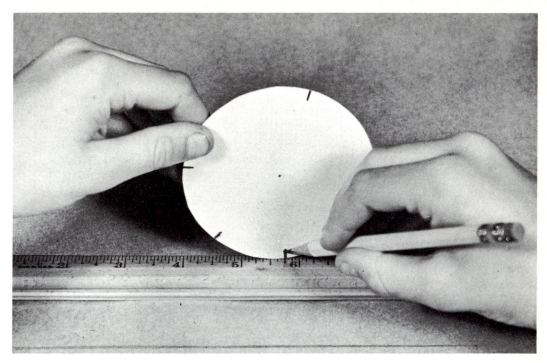

Mark off the disc as you roll it along the ruler.

The coloured sections of the disc . . .

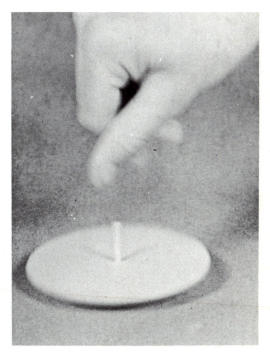

. . . blend into white when it spins.

SEEING THINGS
IN COLOURED LIGHT

In daylight, you see all things in their true colours. How would they appear in coloured light? To find out, you do not have to shine coloured lights on them. All you need do is look at them through pieces of coloured glass or cellophane.

Get two pieces of coloured cellophane, one red and one green. Put a tomato and a cucumber on the table and look at them through the red cellophane. The tomato will seem pale and the cucumber dark. That is because the tomato reflects mainly red light, which can easily go through red cellophane. The cucumber reflects mainly green light, which is stopped by red cellophane, and so it looks dark.

Now look through the green cellophane. This time the result is just the opposite. The cucumber looks pale and the tomato looks dark.

For another experiment, get crayons that are about the same colour as the two pieces of cellophane. Print your name lightly on a piece of paper, making some letters green and some red. Look at the printing through the red cellophane. The red letters will disappear, while the others will show up dark. Now look through the green cellophane and see what happens. Each colour of cellophane fades out the letters that match it.

These experiments show that when you judge the colour of anything, you must allow for the kind of light that falls on it. This is why it is difficult or impossible to match colours by artificial light in a shop.

Through red cellophane, the tomatoes look pale and the cucumbers dark.

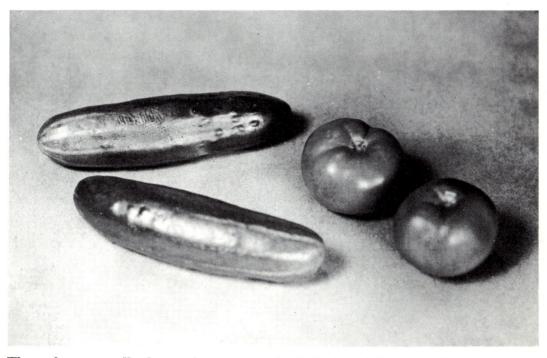

Through green cellophane, the tomatoes look dark and the cucumbers pale.

SCATTERED LIGHT

Why does the sky look blue? Why is the sun orange-red, just before it sets? You can find the answers to these questions by doing an experiment.

Set up a table in bright sunlight coming through a window. Get the glass baking dish you used on pages 22 and 40, put it on the table and fill it with water. Cut a 25 mm (or 1 in.) hole in a large card and set it up in front of the dish. The hole should be level with the water.

Place a mirror at the proper angle to reflect sunlight through the hole and through the water to the far end of the dish. The beam of light may be very weak, but you can see where it comes out of the dish.

Now make the water cloudy by swishing the end of a bar of soap around in it. This will make the beam show up clearly, just like a searchlight in the air (page 2).

Look into the water through the side of the dish. The beam has a bluish colour. Then look through one end of the dish, as in the picture. The beam looks orange-red.

Here is the reason for the colours you saw: whenever a beam of light goes through a haze of any kind, some of the light is scattered off to the sides. The short, blue waves are scattered easily, while most of the longer red ones go straight on. In your experiment, the soap made the water cloudy. This scattered the light and made the beam look bluish from the side and reddish from the end.

The same thing happens to sunlight in the air, where dust and even the air molecules themselves do the scattering. When you look upwards, you see the short waves scattered from the beams of sunlight passing overhead, as in the first drawing. This makes the sky look blue.

When you look at the low sun, the light comes directly to you through many kilometres of air, as the second drawing shows. The shorter waves are scattered aside, and only the longer red ones reach your eyes. This gives the sun a red appearance instead of its usual white glare.

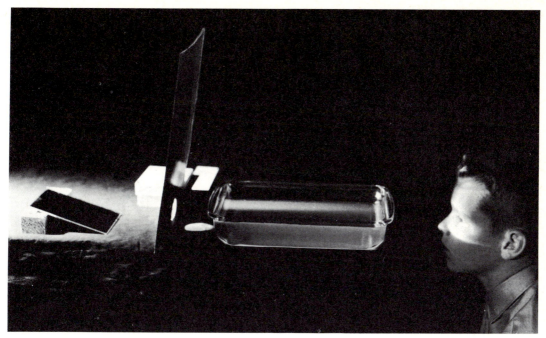

The beam shows up clearly in the water.

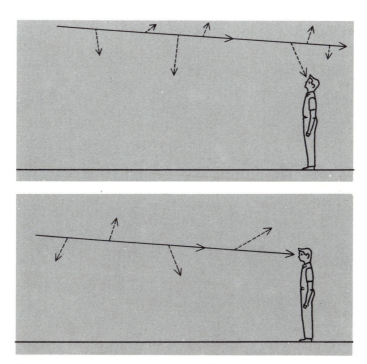

Rays of sunlight are scattered by the air.

LIGHT WAVES CAN STOP EACH OTHER

Dip the open end of a small bottle into soapy water or into some bubble liquid that you can get at a toy shop. A film like part of a soap bubble forms across the open end of the bottle. Tip it forwards until you can see the bright sky (not direct sunlight) reflected in the film. As you watch, you will see coloured bands appear across the film like a row of rainbows.

Where do these colours come from? The soap film has no colour of its own, and the colours do not come from refraction, as in a prism. They are formed by two sets of light waves reflected from the thin film. One set is reflected from the front surface and the other from the back.

Some times these two sets of waves happen to cancel each other as they come out. This leaves only the other colours to be reflected to your eye. Scientists call this the INTERFERENCE of light.

The soap film in your experiment keeps getting thicker towards the lower edge as the water drains down. There are places where the film is just the right thickness for waves of red light to interfere with each other. At those places, what you see is daylight with the red missing, so the soap film looks blue-green there. At other places, where the thickness is different other colours will be cut out.

Interference can be used to cut down the glare of light reflected from glass. This is done by putting on a special coating only a few millionths of a centimetre thick. The picture shows how you can see clearly through the part of the glass that has been coated. Coatings of this kind are put on camera lenses to make the images clearer.

Tip the bottle until the soap film reflects the sky.

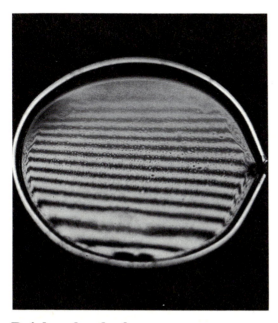

Rainbow bands show up.

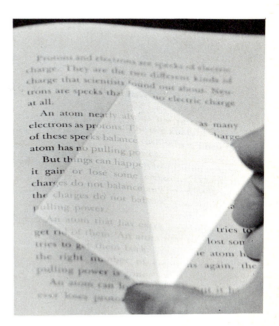

The coating cuts down the glare.

USEFUL BANDS OF COLOUR

Try another experiment that shows interference stripes.

Sprinkle a little talcum powder on a tuft of cotton wool and dab it lightly and evenly all over a small mirror. Darken the room and lay the mirror on the floor right in front of you. Switch on your torch and hold it at your forehead, pointing down at the mirror. Look carefully and you will see coloured stripes all across the mirror.

The same thing happens here as with the soap film in the last experiment. Bands of colour are caused by the interference of two sets of light waves. One set is reflected from the silvered back of the mirror and the other from the powdered front of the glass.

There are other places where interference bands show up. Look at a real pearl button or the inside of an oyster shell and notice the delicate colours. Pearly materials such as these are made up of many thin layers, with films of air between. Some of the waves of light reflected from these films of air interfere with each other. The remaining mixture of waves looks coloured even though the material has no colour of its own.

Scientists find the interference of light useful in many ways. They often need to make a piece of glass that is very accurately flat. They check this by laying it down on another piece that they already know is truly flat. Light shining on the two pieces will make interference bands show up. If these bands are straight and evenly-spaced, as in the picture, the scientist knows that the new piece is as flat as it should be.

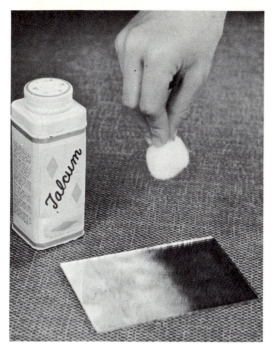

Dust the mirror lightly.

Coloured stripes show up in the mirror.

A light test for flatness

COLOURED SHADOWS

At the very beginning of this book you found out that light travels in straight lines so long as it stays in one material. Then you found that light changes its direction when it bounces back from a surface or when it goes from one material into another.

There is one more way in which light can change its direction. Scientists discovered that when light passes close to the edge of an object or goes through very small openings, some of the waves bend aside. This bending is called the DIFFRACTION of light.

The change in direction is usually too small to notice, but you can do an experiment that shows it plainly.

Tape a pin and a needle to your magnifying lens, as in the picture at the lower left. Turn out all the lights in the room. Switch on your torch and place it on a table.

Sit at another table across the room. Look at the torchlight through the glass with one eye, keeping the other eye covered with your hand. The side of the lens with the pin and needle on it should be towards the light.

Brace your elbow on the table so that your arm is steady, and move the glass slightly back and forth until it seems filled with light.

The picture at the lower right shows what you see when everything is properly lined up. The shadows of the pin and needle are broken up by coloured streaks of light, and they are coloured shadow bands all round the edges.

The streaks and bands are caused by diffracted light. This is light that bends around the sides of things.

Scientists know that light could not bend in just this way unless it travelled in the form of waves. By doing diffraction experiments, they were able to prove that light really is made up of waves.

Arrangement for a diffraction experiment.

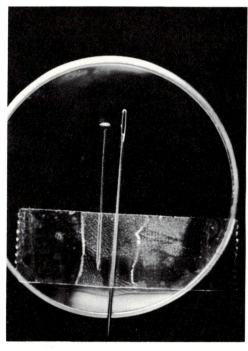

The pin and needle . . .

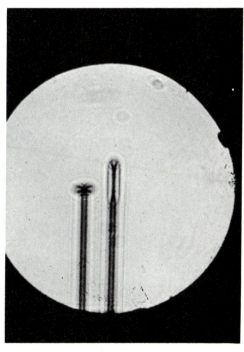

. . . have streaky shadows.

GRATINGS AND WHAT THEY CAN DO

In the last experiment, you saw coloured bands when light was diffracted by the needle and pin. It is much easier to see the bands if you use a long row of evenly-spaced bars instead of just one needle or pin. A set of bars of this kind is called a diffraction GRATING.

Keeping one eye covered, look at a distant light through the outer edge of a chicken feather. Notice how the light coming through the feather is spread out into a spectrum.

Next, face a distant light and hold a gramophone record just below the level of your eyes, as in the picture. Tilt the record until you see the light reflected from the grooves. Each patch of light will show up as a spectrum of colour.

In the feather test, the light had to go through the evenly-spaced openings in the feather. This set of openings formed a diffraction grating. When the different wave lengths of light went through it, they were bent in different amounts to make a spectrum. In the record experiment, the same thing happened when the light was reflected from the sides of the evenly-spaced grooves.

The gratings used in science laboratories have to be made very carefully and exactly by a special machine that makes thousands of scratches, side by side, on a mirror. Light is reflected from the evenly-spaced clear places between the scratches. This forms the spectrum.

A feather has about 27 openings in the space of a centimetre, and a long-playing gramophone record has 120 grooves to a centimetre. A scientific diffraction grating may have 20,000 or more evenly-spaced scratches to a centimetre!

A gramophone record acts like a diffraction grating . . .

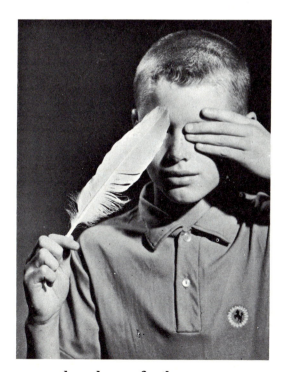

. . . and so does a feather.

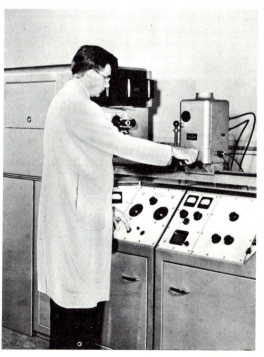

An instrument that breaks up light.

COLOURS THAT
ARE NOT THERE

Seeing colours is one of the most interesting things your eyes do. Light comes into the eye to form an image, and a bundle of more than half a million nerves then carries the message to your brain. The special nerves that see colours get tired when you stare steadily at anything. This can have some interesting results, as an experiment will show.

Close the blinds in a room so that the light is quite dim. For this experiment you will need a yellow lamp bulb. Put it into an unshaded table lamp. Have a large piece of white cardboard ready.

Switch on the lamp and set it about one metre (or 3 to 4 ft) away from you, as in the upper picture. Now, without moving your head, stare steadily at the lighted bulb. After about 10 or 15 seconds, its colour seems to change to blue-white.

Moving as little as possible, switch off the lamp. Then hold up the cardboard and stare at it. After a few seconds you begin to see a bright red image of the bulb, while the rest of the card looks bluish. A picture that looms up in the eyes in this way is called an AFTER-IMAGE.

Keep staring at the card. A bright blue-green colour starts to creep in from the edge of the red image, and the rest of the card now looks pink. You will find these colours more dazzling and beautiful than anything you can imagine.

Blinking or moving your eyes will make the image disappear, but it soon comes back. Even several minutes later, after the room is again fully lighted, you may still see a weak patch of green.

It is hard to believe that all these brilliant colours are only in your eyes. They show up because of the way the nerve endings tire out and come back to normal again.

In order not to strain your eyes, wait about half an hour before repeating the test. This kind of experiment should not be done more than two or three times in one day.

After staring at the lighted lamp . . .

. . . he sees brilliant colours on the blank card.

BUMPY SHADOWS

This experiment must be done in a room that has Venetian blinds on the windows. Wait until the sun comes directly into the room. Then lay a large white sheet of paper or cardboard on the floor. Turn the blind until the shadows of the slats lie in clear stripes on the paper.

Get a long ruler or a curtain rod. Stand a few metres away from the window and hold the ruler so that its shadow slants across the shadows of the slats. You will see large bumps where the shadows cross.

See what happens when you hold out your hand with the fingers spread out so that their shadows cross the shadows of the slats. Also, try holding one finger level and moving it slowly until its shadow comes close to the shadow of a slat.

The strange bumps in the shadows are caused by diffraction. The filling-in or melting-together of shadows happens whenever the two objects that cast the shadows are at different distances from the paper.